SCIENTIFIC AMERICAN EDUCATIONAL PUBLISHING

OUT OF THIS WORLD

10 FUN EXPERIMENTS ABOUT EARTH AND SPACE SCIENCE

BRING SCIENCE HOME

Published in 2025 by The Rosen Publishing Group, Inc.
2544 Clinton Street, Buffalo, NY 14224

Contains material from Scientific American, a division of Springer Nature America, Inc., reprinted by permission as well as original material from The Rosen Publishing Group.

First Edition

Editor: Kristen Rajczak Nelson
Designer: Rachel Rising

All illustrations by Continuum Content Solutions

Photo Credits: pp. 3, 4, 9, 12, 27, 30, 33, 37, 45, 48, 51, 54 cve iv/Shutterstock.com; pp. 4, 5, 9, 15, 21, 27, 33, 39, 45, 51, 56 Anna Frajtova/Shutterstock.com.

Activity on p. 5 by Science Buddies, Ben Finio (December 7, 2017): p. 9 by Science Buddies (October 8, 2015); p. 15 by Science Buddies (August 7, 2014); p. 21 by Science Buddies (February 20, 2014); p. 27 by Science Buddies (February 6, 2014); p. 33 by Science Buddies (August 8, 2013); p. 39 by Science Buddies, Svenja Lohner (September 27, 2018); p. 45 by Science Buddies, Svenja Lohner (August 2, 2018); p. 51 by Science Buddies, Sabine De Brabandere (July 12, 2018); p. 56 by Science Buddies, Megan Arnett (March 15, 2018).

Names: Scientific American Editors (New York, New York), editor.
Title: Out of this world : 10 fun experiments about earth and space science / Scientific American Editors.
Other titles: Bring science home.
Description: Buffalo, New York : Scientific American Educational Publishing, an imprint of Rosen Publishing, [2025] | Series: Bring science home | Includes bibliographical references and index.
Identifiers: LCCN 2024005928 (print) | LCCN 2024005929 (ebook) | ISBN 9781725350618 (library binding) | ISBN 9781725350601 (paperback) | ISBN 9781725350625 (ebook)
Subjects: LCSH: Space sciences--Experiments--Juvenile literature. | Astronomy--Experiments--Juvenile literature. | Earth sciences--Experiments--Juvenile literature.
Classification: LCC QB500.264 .O98 2025 (print) | LCC QB500.264 (ebook) | DDC 520.78--dc23/eng/20240216
LC record available at https://lccn.loc.gov/2024005928
LC ebook record available at https://lccn.loc.gov/2024005929
Manufactured in the United States of America

CPSIA Compliance Information: Batch #CSSA25. For further information contact Rosen Publishing, New York, New York at 1-800-237-9932.

CONTENTS

Introduction 4

Make a Black Hole 5

Waves in Slow Motion ⚛ 9

Starry Science: Measuring Astronomical
Distances Using Parallax 15

Seasonal Science: The Reason for the Seasons 21

Sliding Science: How Are Landslides Caused? ⚛ 27

Shooting Star Science: Craters and Meteorites ⚛ . . . 33

Measure the Pressure 39

Create a Sea Breeze ⚛ 45

Warm in the Sun ⚛ 51

It's About Time ... to Make a Sundial! 56

The Scientific Method 61

Glossary 62

Additional Resources 63

Index 64

⚛ THESE ACTIVITIES INCLUDE SCIENCE FAIR PROJECT IDEAS.

INTRODUCTION

From outer space to your own backyard—science fun is all around us! In these experiments, you'll create a black hole without going to space, find out the reason for the different seasons, create a sea breeze without having to take a vacation, and more—all while finding out more about Earth and space science!

Projects marked with include a section called Science Fair Project Ideas. These ideas can help you develop your own original science fair project. Science fair judges tend to reward creative thought and imagination, and it helps if you are really interested in your project. You will also need to follow the scientific method. See page 61 for more information about that.

Make a Black Hole

GET SUCKED INTO SCIENCE WITH THIS FUN PHYSICS ACTIVITY YOU CAN DO AT HOME!

PROJECT TIME

15 to 20 minutes

KEY CONCEPTS

Physics
Mass
Gravity
Space-time

Do the words "general relativity," "gravity well," and "space-time continuum" sound intimidating? Don't worry, you don't have to be Albert Einstein to understand them! Try this fun activity to learn about these concepts and black holes—using some common household materials.

BACKGROUND

We experience gravity all the time. It's such a regular part of our lives that we might not give it much thought. It keeps our feet planted on the ground so we don't float off into space. If we throw a ball up, it comes back down. Pretty easy, right?

In classical, or Newtonian, mechanics, gravity is described as a force that acts between two objects, pulling them together. Technically any object with mass exerts a gravitational force on other objects. This force between objects, however, is usually too small for us to feel. You don't feel a sideways gravitational pull from the person standing next to you, for example, or even from bigger objects such as cars or buildings. We only feel the gravitational force from huge masses, such as Earth.

Einstein's theory of general relativity describes gravity a little differently. First it describes space and time as a "fabric," or "continuum" called space-time. (If you think of space as three-dimensional and you add time as another dimension, then you get something that has four dimensions.) General relativity describes gravity as the curvature of this four-dimensional space-time, which is caused by mass. So large masses such as Earth cause a large curvature in space-time whereas smaller masses such as our bodies barely cause any curvature. If that sounds confusing, don't worry—this activity will help you visualize it.

Although Earth's gravity is very strong, we can build rockets and spaceships that can escape it and fly off to the rest of the solar system. We can also send electromagnetic radiation such as light or radio waves out into space. A black hole, however, is an object in space with so much mass and with gravity so strong (that is, it curves space-time so much) that almost nothing can escape it—not even light. Anything that passes too close to a black hole (inside a region called the event horizon) will be sucked in. You might wonder how astronomers know black holes exist. After all, if light cannot escape them (meaning they do not emit light or reflect light), how can we see them? Scientists know black holes exist because we can observe their effects on other celestial bodies such as stars. For example, a star's orbit may be affected by the gravity of a nearby black hole, and we can observe matter being sucked into a black hole (before it disappears behind the event horizon).

In this activity, you will use some simple household materials to demonstrate how a large mass such as a black hole curves space-time—and what effect this has on other nearby objects.

MATERIALS

- Stretchy fabric, such as a polyester/Lycra T-shirt (not cotton) or a large athletic bandage

- Round, heavy object, such as an apple, orange, or pool ball
- Two marbles or Ping-Pong balls
- Two helpers
- Marker and ruler (optional)

PREPARATION

- If you are using an athletic bandage, cut a large rectangular piece out of it.
- Have your two volunteers hold your stretchy fabric by the corners, pull it taut, and hold it flat. This fabric represents space-time.

PROCEDURE

- Put a marble somewhere on the fabric and watch it. *What happens?*
- Now roll a marble across the fabric and watch how it moves. *What shape does its path take?*
- Place your heavy, round object in the middle of the fabric. Make sure your volunteers still pull the fabric tightly enough that it does not have any wrinkles or bumps in it—try to make sure it's as smooth as possible. *What happens to the fabric?*
- Next place the marble near the edge of the fabric and let it go. *What happens?*
- Now try to roll the marble from one side of the fabric to the other. *Is it difficult? What shape does the marble's path form?*
- Try rolling the marble across the fabric at different speeds. *What happens if you roll the marble slowly? Faster? How does the shape of its path change? Is there a minimum speed you need to roll the marble to prevent it from getting sucked in toward the larger ball?*
- Try rolling two marbles on the fabric at once. *Do the marbles appear to affect each other's motion (assuming they do not collide)?*

EXTRA

If you don't mind drawing on your fabric, lay it flat on a table. Use a ruler and marker to draw a grid of squares on the fabric. The grid should consist of lines that intersect each other at right angles. Now repeat the activity. *What happens to the grid?*

OBSERVATIONS AND RESULTS

You might think the results of this activity are obvious at first. When you hold the fabric flat and put a marble on it the marble doesn't move. If you roll the marble across the flat fabric, it moves in a straight line. No surprise there! Remember, however, that your fabric represents space-time. When there is no large mass present on the fabric, it doesn't curve at all, which is why your marble can just sit there or move in a straight line when you roll it. This is how a beam of light would travel through space if there were no black holes nearby.

When you place a heavy ball in the middle of your fabric, it causes the fabric to curve downward, just like the presence of a large mass (such as a planet, star, or black hole) causes space-time to curve, forming a gravity well. It's now nearly impossible to get the marble to hold still—it always rolls toward the middle. When you try to roll the marble in a line, it follows a curved path. As long as the marble doesn't get too close to the middle you could probably still roll it from one end of the fabric to the other—similar to how a beam of light that passes near a black hole (but not inside the event horizon) will curve but then keep going. If it gets too close to the center, however, the marble gets sucked in and cannot escape—just like light cannot escape if it ends up inside a black hole's event horizon.

When you rolled two marbles at once you probably noticed that (as long as they didn't collide) they did not really affect each other's motion. This is because the marbles are not heavy enough (they do not have enough mass) to introduce any additional curvature in the fabric. All of the curvature is due to the much heavier ball in the middle. This demonstrates why we only notice Earth's gravity pulling us down and we don't get pulled sideway by other objects such as people, cars, or buildings—their gravity is much too weak compared to Earth's.

CLEANUP

Return all materials to the places where you found them. Throw away any garbage.

Waves in Slow Motion

HOW DO WAVES WORK? YOU WILL BE ABLE TO SEE FOR YOURSELF WITH THIS FUN, SLOW-MOTION SCIENCE ACTIVITY. CATCH A WAVE!

PROJECT TIME

15 to 20 minutes

KEY CONCEPTS

Wave
Pressure
Density

Have you ever noticed seagulls bobbing up and down on the ocean? You might have also seen surfers catch a wave that takes them to shore. Maybe you have floated on a lake, going up and down as a wave passed by. Or perhaps you have seen debris, such as driftwood, that has been washed up by waves. Water waves are fascinating—they come in all sizes, from a tiny ripple to monster waves that are more than 30 feet (9 m) high. You have seen them but do you know what drives them—and how they move across oceans and seas? In this activity, you will bring the ocean home and make waves in a bottle. You will also sharpen your observation skills and find out why some waves are slow and others are fast.

BACKGROUND

Wind causes much of the waves on seas and oceans. Here's how it works: Wind causes water to pile up above the undisturbed water level. The water under the crest (or peak) of a wave gets the extra water pushing down on it. The extra weight then pushes water out from under the peak of the wave. The moving water overshoots and creates a peak in a new place, and so the wave moves on.

Oceans and seas are layers of water under a layer of air, the atmosphere. Because air is much less dense than water (weighing about 1,000 times less than the same quantity of water), it seems like air would have little influence on a wave. So in order to understand how waves work, you will add a layer of oil on top of the water. Curious about how this relates to air and affects a wave on the water surface? Roll up your sleeves and try this rocking science activity!

MATERIALS

- Two similar, clean, transparent bottles with caps (2-liter soda bottles work well)
- Vegetable oil, enough to fill one bottle two-thirds full
- Water
- Food coloring (Blue is a good choice.)
- Towel
- Pen and paper
- A workspace and covering material (such as an old towel) where spilled oil and water can easily be cleaned up
- Funnel for easier filling of the bottles (optional)
- Two small pieces of wax (Pieces cut off a candle work well. Have an adult help you safely cut these.) (optional)

PREPARATION

- Place a towel (or other protective layer that can get dirty) on your work area.
- Fill both bottles about one-third full of water. A funnel can help you pour water into the bottles. Add a few drops of food coloring.

- Close one bottle and set it aside.
- Gently fill up the other bottle with vegetable oil. Here again, a funnel can prevent spills.
- Close the bottle and set it aside.
- Check that both lids are screwed on well.
- Move the prepared bottles carefully to the area where you will make observations.
- Let the bottles rest on their sides until you are ready to observe.
- *What can you see in the bottle containing water and oil? Is the water floating on top of the oil, the oil on top of the water, or are both liquids mixed? Which one do you think is denser, the oil or the water?*
- *What do you think is on top of the water in the bottle containing only colored water? Is there anything or is it empty space?*

PROCEDURE

- Start with the bottle containing only water. Hold this bottle horizontally and lift one end to cause a wave. Once the wave has started, bring your bottle back to a horizontal position and observe the wave.
- Repeat with the bottle containing oil and water. *Do you notice a difference in the waves produced? Are the waves in one bottle moving faster than in the other? Are they higher?*
- To make the comparison easier, count each time the wave bounces back on the cap or bottom side of the bottle. Start with the bottle containing only water. Hold the bottle horizontally and lift one side to create the wave. Bring your bottle back to the horizontal position and keep count as the wave reaches the other side of the bottle.

- Add a number each time the wave reaches the cap or the bottom side of the bottle. Do this until the wave has died out.
- Repeat the previous step with the other bottle. *Did you notice a difference? Did you need to count more slowly for one bottle than the other? What does that indicate and why could that be?* The extra activities below might give you a clue.

SCIENCE FAIR IDEA

Open the bottles, put a little piece of wax inside both bottles and close them again. Because wax is denser than oil but less dense than water it will float on water but sink in oil. You might need to wait a little while for the wax to sink through the oil until it reaches its place on top of the water. Think of the wax as a seagull, a boat, or a surfer floating on top of the water. Create a wave like you did before but now watch how the wax moves. *Does it move back and forth, or rather up and down with the wave? Can you explain your observations?*

EXTRA

If you have a stopwatch, you can measure the wave's period. This is the time it takes for one wave to pass by—from crest to trough and back to crest. *Looking back at the observations you made earlier, do you think the period is the same for waves with air on top and waves with oil on top?* Use your stopwatch to time how long it takes for your wax object to start at the top of a crest, go down into a trough and back up to the top of a crest. This time measured is exactly the period of the wave. *Is the period the same for waves with air on top and waves with oil on top?*

EXTRA

The frequency of a wave is how many waves pass by in a second. *Looking back at the observations you made earlier, do you think the frequency is the same for waves with air on top and waves with oil on top? Can you find a way to measure or calculate the frequency of the waves you created?*

OBSERVATIONS AND RESULTS

Did you notice that water with oil on top produces surprisingly slow waves? A wave moves because the extra weight of the liquid in the wave's peak pushes water from under the peak to places where the water is shallower. This happens in both bottles.

For the bottle with water and air, both the water and the air push on the deeper layers of water. Air is about 1,000 times less dense than water, so when a wave passes by, deeper layers of water mainly get the weight of all the water piled up in the wave. The difference in weight between the crest and the trough (or lowest point) is the wave's driving force. In this scenario, the big difference in weight results in a fast-moving wave.

Similarly, for the bottle with water and oil, water and oil push on the deeper layers of water. Just like air, oil is less dense than water (about two-thirds as dense), so an amount of oil weighs less than the same amount of water. Thus, as water piles up in a wave, more water and less oil will weigh down on a spot underneath the wave peak. At the shallowest point of a wave, however, more oil and less water weighs down on the water below. As a result, the difference in weight of the liquids at a point under the crest and a point under the trough (or lowest point of the wave) is much smaller in this scenario, so the driving force of the wave is much smaller, although it still forces water to move from underneath the wave peak. In addition, the water and the heavier-than-air oil both have to be moved, so the liquids flow much more slowly than when the light air flows on top of water. What you observe is a wave on the water surface that travels much more slowly.

If you did the extra activity, you probably observed that the wax mainly moved up and down while the wave moved forward and backward. This is because in a wave it is the water underneath the wave that moves horizontally; the water on the surface moves up and down.

CLEANUP

You can pour the contents of the bottle with only colored water down the drain. But do not pour oil down the drain!

Set your bottle with water and oil upright and give the liquids some time to separate. Select a container or bottle (such as the one you used for colored water) in which to collect your oil. Pour the layer of oil into a clean container or bottle. It is okay if some water flows over as well, just do your best to transfer most of the oil. (Use a funnel if one is available.) You can pour the remaining colored water down the drain. You can throw your sealed bottle of oil in the trash, compost it or search online for a place to donate your used cooking oil. (Used cooking oil can be turned into biodiesel, a nontoxic and biodegradable fuel that can power cars and buses and produces fewer air pollutants than petroleum-based diesel.)

Starry Science

Measure Astronomical Distances Using Parallax

HOW FAR AWAY ARE THOSE STARS? FIND OUT IN THIS FUN ACTIVITY!

PROJECT TIME

25 to 35 minutes

KEY CONCEPTS

Astronomy
Stars
Physics
Measurement

Do you enjoy going stargazing on a warm night? Summer can be a great time to watch the stars as well as other celestial events, such as the impressive Perseids meteor shower that happens annually. Did you know that ancient astronomers could actually measure the distance from Earth to faraway stars? How could they do this without modern technology? In this activity, you will find out by exploring the relationship between the distance of an object and the viewing perspective (also known as parallax), which can be used to measure how far away distant stars are.

BACKGROUND

How do astronomers know how far planets, stars, and galaxies are from us? They use a visual phenomenon called parallax to measure stellar distances. Parallax is the way an object appears to move, as it appears to change position when it is seen from two different locations, or perspectives.

To see parallax for yourself, hold out your arm and stick up your thumb. Closing one eye, line up your thumb with an object across the room. Now quickly switch your eyes while keeping your thumb in the same position. (You can use your free hand to cover your other eye if necessary.) Notice that the object you were looking at is no longer lined up with your thumb. Don't believe what you're seeing? Repeat again starting with the other eye. This optical illusion is because of parallax. The difference in distance between your two eyes makes your thumb (a relatively nearby object) line up differently with the object that is across the room (a relatively distant object).

When a person looks at stars when Earth is at different positions in its orbit, closer stars will appear to change position much more relative to stars that are farther away. This apparent movement, or parallax, can be used to help figure out distances between Earth and specific stars.

MATERIALS

- A wide-open space (It could be your backyard or a park.)
- Two Hula-Hoops (Alternatively, you could use two flat rocks or bricks that you can sit on.)
- A yardstick or meterstick (Use one that has clear markings so that it can be read from a distance.)
- Small table or bar stool (This will need to be able to be taken outside.)
- A thick rubber band
- A large rock (It should be at least the size of a baseball, but not so large the rubber band cannot fit around it.)
- Measuring tape
- A scratch piece of paper and a pen or pencil (optional)

PREPARATION

- If you are using a scratch piece of paper, draw three columns on it and label them "Left," "Right," and "Difference."
- Take your Hula-Hoops (or flat rocks), small table (or bar stool), rubber band, large rock, and measuring tape outside to a wide-open space, such as a park or your backyard.
- Find a distant object—ideally one that is tall and narrow, such as a tree, light pole, or post. This will be the "distant object" you use in this activity.
- Giving as much distance between you and the distant object as possible (at least 20 to 30 steps, but farther is even better), face the object and place the two Hula-Hoops on the ground, one on your right and one on your left, so that they are nearly touching. (If you are using flat rocks instead, place them so that they are about 3 feet [0.9 m] apart, to your left and right as you face the distant object.) Each Hula-Hoop (or flat rock) will be an "observation point."
- Walking from the Hula-Hoops toward the distant object, place the small table about three to five steps away from the edge of the Hula-Hoops (or about five to seven steps away from the flat rocks).
- On top of the table, place a large rock. Loop the rubber band around the rock and yardstick (or meterstick) to hold the yardstick against the rock so that the yardstick is horizontal (running to your left and right) and its markings are facing you (toward the Hula-Hoops). Center the yardstick along the rubber band. The table (with the rock, yardstick, and rubber band) will be the "near object" you investigate.
- Your setup should now be ready for you to do some testing! But before you do, make sure that all of the parts of your setup are lined up well. Specifically, make sure that the space between the two Hula-Hoops, the rubber band, and the distant object all roughly line up along an imaginary line. If needed, you can shift the table to the left or right to make the rubber band line up.

PROCEDURE

- Sit in the center of the left Hula-Hoop (or on the flat rock on the left) and look toward the distant object. *Which number on the yardstick does the distant object appear to line up with?* If you are using a scratch piece of paper, write this number down in the "Left" column.

- Sit in the center of the right Hula-Hoop (or on the flat rock on the right) and look toward the distant object. *Which number on the yardstick does the distant object now appear to line up with?* If you are using a scratch piece of paper, write this number down in the "Right" column.

- Now move the table (with the rock and yardstick on it) forward another three to five steps (so that it is closer to the distant object).

- Sit in each Hula-Hoop again and see which number on the yardstick the distant object appears to line up with now. If you are using a scratch piece of paper, write down your results. *Do you see a difference in where the distant object appears to line up on the yardstick compared with when the table was closer to you?*

- Repeat this process at least three more times (each time moving the table a little closer to the distant object and then looking at how the distant object lines up on the yardstick from your Hula-Hoop vantage points). *How does where the distant object lines up on the yardstick appear to change as the near object (the table with the yardstick) is moved closer and closer to the distant object?*

- If you used a scratch piece of paper to record your results, you can subtract the number in one column from the other and write your result down in the "Difference" column for each distance you tested. *Do you see a relationship between the distance (between the observation points and the near object) and the difference in the measurements you took from the left and right perspectives?*

- ***What do you think your results tell you about how astronomers use parallax to measure how far away a relatively nearby star is?***

EXTRA

In this activity, you moved the near object (the table with the rock, yardstick, and rubber band) different distances away from the observation points (the Hula-Hoops). Another factor for measuring parallax is the distance between two observation points. *Can you think of a similar activity you could do to test this variable? How does the distance between observation points affect parallax?*

EXTRA

If you watch your favorite constellation over several nights, you will notice that the stars move together as a group. Compare the movement of the constellation to nearer objects, like the moon or a planet. *Which objects move more quickly? Can you pick out the difference between planets and stars using this method?*

EXTRA

Parallax is similar to the process our brains use when calculating depth perception. Intuitively, you know which objects are near and which are far. You could try testing depth perception by comparing binocular vision and monocular vision. *How does each affect depth perception?*

OBSERVATIONS AND RESULTS

As the near object moved farther from the observation points, did the apparent movement of the object decrease (as measured from the left and right observation points)?

When a relatively distant object is viewed from two different points, it appears to move less compared with a relatively nearby object. Similarly, in this activity you should have seen that as the "near object" (the table with the rock, yardstick, and rubber band on it) got closer to the "distant object" (the tree, light pole, etcetera), it appeared to move less (when you compared its apparent position between the left and right "observation points" inside the Hula-Hoops). It might have been hard to tell the difference for the first two measurements, but the relationship should have become clearer after that point.

Because the apparent movement of an object (the parallax) depends on how far the object is from the observation points, astronomers can figure out how far away relatively nearby stars are. This is done by looking at a nearby star's apparent movement relative to distant stars when they are all viewed from different observation points (that is, from different points in Earth's orbit around the sun).

CLEANUP

Return all materials to the places where you found them.

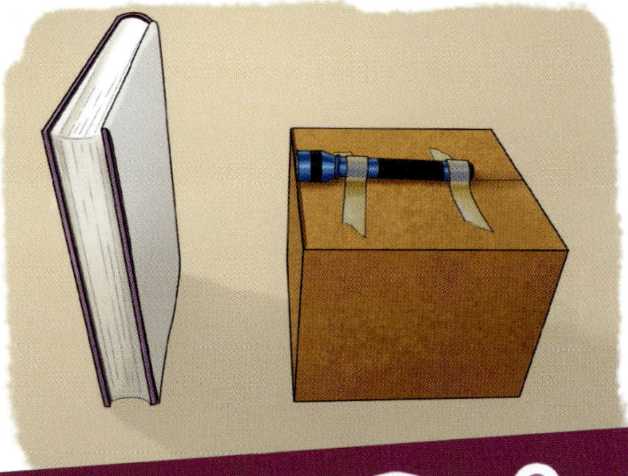

Seasonal Science
The Reasons for the Seasons

WINTER, SPRING, SUMMER, AND FALL:
FIND OUT WHY WE HAVE THEM ALL!

Have you ever lived somewhere where you get to experience the full glory of all four seasons? If so, you know well the full blossoms and dramatic skies of spring; the long, sun-drenched days of summer; the trees shaking in crimson and gold in fall; and the sparkling snows of winter. But do you know why we have these seasons over and over again in a cycle as predictable as sunrise and sunset? It actually has to do with Earth's tilt. In this science activity, you'll investigate how this tilt affects how the sun's rays strike our planet and create seasons.

PROJECT TIME

15 to 20 minutes

KEY CONCEPTS

Astronomy
The sun
Light
The seasons
Earth's orbit

BACKGROUND

In Earth's Northern Hemisphere, summers are hot and filled with many hours of strong sunlight whereas winters are cold due to shortened daylight hours and weak sunlight. Why is this? One big part of the answer is that Earth is tilted on its axis. To visualize this axis, picture an imaginary stick piercing Earth from its North to South Poles. Earth spins once around this axis every 24 hours. While spinning like this, our planet also circles the sun in a big orbit, completing this loop in about 365 days.

This axis isn't straight up and down as Earth orbits around the sun. Instead, it is tilted at approximately 23 degrees and also remains fixed, always aligned in the same direction in space. This tilt changes how the sunlight hits Earth at a given location in its yearly orbit. When it's summer in the Northern Hemisphere, the top part of the axis (the North Pole) points more toward the sun, and the sun's rays shine more directly on the Northern Hemisphere (where the continents of North America, Europe, and Asia as well as the northern parts of Africa and South America are located); at the same time in the Southern Hemisphere (Antarctica, Australia, most of South America, and the southern third of Africa), where it's winter, the South Pole end of the axis is tipped away from the sun, and its rays hit that half of Earth on a slant.

MATERIALS

- Small cardboard box, stepping-stool, brick, or large block of wood
- Flashlight
- Masking tape
- Large, firm book or cutting board
- Sheet of paper
- Pen or pencil
- Helper (optional)

PREPARATION

- If you are using a cardboard box, tape it so it is sealed shut.
- Place the sealed-shut cardboard box, stepping-stool, and brick or block of wood on a table or on the floor.

- Lay the flashlight on its side on top of the cardboard box (or other object). Line up the light-emitting end of the flashlight with the edge of the box. Use masking tape to tape the flashlight down so it can't roll around. The flashlight will represent the sun.

- Tape a sheet of paper to the book (or cutting board) so that the paper will be stiff enough to tilt and so that you can draw on it. This paper will represent part of Earth's surface.

PROCEDURE

- Put the book (with the taped-on paper) vertically in front of the flashlight and move the book closer or farther away from the flashlight until the light on the paper forms a medium-size, sharp circle that is about 2 to 3 inches (5 to 7.5 cm) in diameter. Make sure that there are at least 2 inches (5 cm) of paper above the top of the circle of light that is shining on the paper. If needed, raise the paper (by taping it higher on the book) and/or lower the flashlight (by taping it to a shorter cardboard box, etcetera).

- While you are holding the book (or having a helper hold it for you) vertically in front of the flashlight, look at the light that is shining on the paper. *How bright is the light on the paper?* Using the pen or pencil, draw around the outline of the light on the paper.

- Without changing the distance between the book and the flashlight, tilt the book back, away from the flashlight by about 45 degrees. In other words, the book should roughly form a 45-degree angle with the floor or tabletop. *How did the light on the sheet of paper change? How bright is the light on the paper now?* Again, draw around the outline of the light on the paper. **Tip:** If the top of the outline goes off the top of the paper, you can tilt the book back toward the flashlight a little until the entire outline fits on the paper.

- If you want, try moving the book back and forth between the vertical position and the 45-degree angle position a few times. Be sure not to change the actual distance between the book and the flashlight. *How does the light on the paper change as you change its tilt angle?*

- Turn off the flashlight and look at your sheet of paper. *How did the outline change when the book was tilted 45 degrees? How did this correlate with a change in light intensity? What degree of tilt do you think is most similar to the light that North America experiences in summer? What about in winter?*

EXTRA

Repeat this activity using a larger range of degrees, such as 20, 30, 40, etcetera. You will want to do this using a protractor at the base of the book as you tilt it. *How does the light outline change as you increase the angle of the book?*

EXTRA

Try repeating this activity, but instead of a blank sheet of paper use graph paper. When you are done making your outlines, you can count how many squares are filled by light when the book isn't tilted compared with when it's at a 45-degree angle. *Using these numbers, just how different are the two outlines in size?*

EXTRA

You could repeat this activity using a light meter, which would let you quantify your brightness observations. *Just how much brighter is the light on the paper at one angle compared with another angle?*

OBSERVATIONS AND RESULTS

Was the light on the paper much brighter when the book was vertically in front of the flashlight compared with when the book was at a 45-degree angle away from the flashlight? Did the outline get bigger and elongated when the book was tilted away from the flashlight?

In this activity, you should have seen that when the flashlight shined on the sheet of paper placed straight in front of it, the light formed a crisp, bright circle. When the book was tilted back 45 degrees, away from the flashlight, it should have made an oval shape that was much dimmer and larger (almost twice the size of the first outline). As the book is tilted away from the flashlight, the light rays hitting the paper's surface become more slanted. Slanted light rays are weaker because they cover a larger area and heat the air and surface less than direct rays do. The same thing happens with Earth and the sun. When Earth's North Pole is tilted toward the sun, the direct rays make the sunlight stronger and thereby warmer in North America—causing it to be summertime—compared with when the North Pole is tilted away from the sun. In that instance, North America gets less direct rays and more slanted ones, causing it to be colder—what we know to be wintertime. The middle of the planet, the equator, never gets tilted too far from the sun, which is why most places closer to the equator, such as Florida, have seasons that are defined less by temperature change than do places farther away from the equator, such as Minneapolis, Minnesota.

CLEANUP

Remove the paper from the book or cutting board. Recycle or reuse any cardboard or paper. Return all other materials to the places where you found them.

Sliding Science
How Are Landslides Caused?

CAN PENNIES IMITATE A LANDSLIDE? FIND OUT IN THIS ACTIVITY!

Have you ever seen a video of a landslide? Landslides are powerful geologic events that happen suddenly and cause devastation in areas with unstable hills, slopes, and cliff sides. Each year in the U.S., landslides can cause great damage to buildings and property, in addition to changing the surrounding habitats. In this science activity, you will model landslides using a clipboard and pennies, and investigate how friction and the angle of a hill's slope affects potential landslides.

PROJECT TIME

20 to 25 minutes

KEY CONCEPTS

Slopes
Angles
Forces
Gravity
Friction

BACKGROUND

A landslide is any geologic process in which gravity causes rock, soil, artificial fill, or a combination of the three to move down a slope. Several things can trigger landslides, including the slow weathering of rocks as well as soil erosion, earthquakes, and volcanic activity.

One major force all landslides have in common is that they are propelled by gravity. We normally think of gravity pulling an object vertically down, such as when you drop a ball straight down. But on a slope, gravity gets slightly more complicated. Any force (such as gravity) has magnitude and direction. On a slope, gravitational effects can be separated into a component that's parallel to the slope (which pulls the object down the slope) and a component perpendicular to it (which pulls the object against the slope's surface). As the angle of the slope increases (making it steeper), gravity's parallel component increases and the perpendicular component decreases, thereby overcoming resistance for downward movement. This resistance is called friction and depends on the perpendicular component of gravity, along with the slope's and object's surfaces. When the parallel component becomes greater than the perpendicular component, the object slides down the slope. In other words, the critical maximum slope from horizontal—called the angle of repose, which is the greatest angle that an object will remain at rest—has been surpassed.

MATERIALS

- Clear tape
- Eight pennies
- Hard clipboard
- Paper towel
- Ruler (optional)

PREPARATION

- Take a piece of tape a little longer than the length of four pennies lined up next to one another (about 3.5 inches [9 cm] long) and set two pennies on the tape so the pennies are touching, side by side.

- Set one penny on each of the two pennies on the tape so that you have two stacks of pennies with two pennies in each stack. Then wrap the tape lengthwise completely around the pennies so that they are held in place, still stacked and side by side. The tape should slightly overlap on the top side.

- Repeat this with the four other pennies so that you have made two taped stacks of pennies like this.
- Cut out a strip of paper towel that is slightly longer than the length of one of the stacks of pennies, and the same width as the pennies. (In other words, the paper towel strip should be about 2 to 2.5 inches [5 to 6.5 cm] long and almost 1 inch [2.5 cm] wide.)
- Take one of the taped penny stacks and make sure the rough, exposed tape edges are on the top (and the smooth side is on the bottom). Then, using two small pieces of tape, tape the paper towel strip lengthwise on to the stack of pennies so that both edges of the strip curve around to the top side and are taped there. Do not put any tape on the bottom side, which should be completely covered only by the paper towel strip.

PROCEDURE

- Set the clipboard on a flat surface. Clip a paper towel sheet onto the clipboard. (If you cut a strip out of the sheet for the penny wrapping, put that space at the bottom of the clipboard.)
- Place both penny stacks you made on the clipboard so that they're both touching the clip at the top. They should be touching the clip lengthwise but not touching one another.
- Make sure both stacks are placed so that their rough tape edges are facing up, and the paper towel strip or smooth taped side of the stacks is down, touching the clipboard. *How does the bottom of each stack feel compared with one another? Is one much smoother than the other?*
- Holding on to the clipboard clip, slowly and steadily lift that end of the clipboard. (Make sure the opposite side stays down, touching the flat surface.) *Which stack of pennies slides down the clipboard first as you increase its angle?* Stop tilting the clipboard as soon as one of the stacks of pennies starts to slide down.

- Repeat this process at least nine more times for a total of 10 trials. Each time be sure to start with the clipboard laying flat on a flat surface and with both stacks of pennies sitting next to one another by the clip. Also make sure to slowly lift the clipboard each time. *For each trial, which stack of pennies slides down the clipboard first? Are your results fairly consistent?*
- *If one stack of pennies usually slid down the clipboard first, why do you think this happened? Why do you think the angle of repose (the angle after which an object slides down a slope) may have been different for the two different stacks of pennies? What do you think your results might have to do with friction?*

EXTRA

You could repeat this activity, but use a protractor to quantify your results. *What is the exact angle at which the different penny stacks start to slide down the slope? How do their angles of repose compare exactly?*

EXTRA

Try this activity again, but this time try making different size penny stacks (with more or fewer pennies) and compare their angles of repose. *How does the size of the penny stack affect how it slides down the slope?*

SCIENCE FAIR IDEA

Grab some other small objects and try repeating this activity. For example, you could make different objects from LEGO. *Do you get similar results? What factors do you think are most important in determining whether an object will first slide at a lower or steeper incline?*

OBSERVATIONS AND RESULTS

Did the tape-only penny stack usually start sliding down the clipboard first when you slowly raised the clipboard, increasing the angle of the slope?

The majority of the time the stack of pennies that were only coated in tape (and not a strip of paper towel) should have started sliding down the clipboard before the other stack of pennies did as the clipboard was raised up by its clip. For example, out of 10 trials, the tape-only penny stack may have started sliding before the paper towel-wrapped stack in all of the trials. The resistance for downward movement on the slope is called friction, and it depends on the component of gravity that is perpendicular to the slope as well as the surfaces of the object and the slope itself. Because there was a greater amount of friction between the two paper towel-coated surfaces rubbing against one another than there was between the paper towel-coated surface and the tape-coated surface, the penny stack with a paper towel strip on it had a greater amount of friction, or resistance to movement, when going down the slope. This greater amount of friction should have given the paper towel-coated stack a greater angle of repose compared with the tape-only stack.

CLEANUP

Remove the pennies from the tape. Throw away any garbage. Return all items back to where you got them.

Shooting Star Science

Craters and Meteorites

CREATE YOUR OWN METEORITE AND SEE WHAT HAPPENS WHEN IT HITS EARTH IN THIS FUN ACTIVITY!

PROJECT TIME

15 to 20 minutes

KEY CONCEPTS

- Astronomy
- Physics
- Meteors and meteorites
- Size
- Geometry

Have you ever seen a "shooting star" race across the sky at night? Shooting stars are meteors zooming at high speed through Earth's atmosphere and burning up along the way from friction. The Perseids, which are impressive meteor showers that happen each year from mid-July to late August, have their peak activity around August 11 to 14. They're caused by Earth traveling through debris from the comet Swift-Tuttle. Luckily, most meteors are tiny and burn up before hitting the ground. But if a meteor doesn't disintegrate, what's its impact on Earth? In this activity, you'll explore how a meteorite's size is related to the size of the crater it makes on impact.

BACKGROUND

Craters are round, bowl-shaped depressions surrounded by a ring. They're made when a meteoroid in space collides with a planet, moon, or other astronomical body. ("Meteorite" is what a meteor is called if it does not burn up before it lands. When a meteor is in space, it is known as a "meteoroid.") Craters are what make our moon look like a ball of Swiss cheese. Each round hole is where a meteorite impacted the surface of the moon, so craters are often called "impact craters." Often, the meteorite that creates the crater explodes on impact, so only the crater is left as an empty reminder of the collision.

Moons and planets have been impacted by meteorites since the formation of our solar system. We see many craters on the moon because it doesn't have much of an atmosphere. An atmosphere (such as the one we have) often causes enough friction to make the falling rocks burn up. The moon, unlike Earth, also lacks weather and water features that eventually erode craters as well as tectonic activity to renew its surface. On Earth, however, there are only about 170 scientifically confirmed impact craters (others are thought to have existed but have been changed beyond recognition over time). And many of these confirmed craters are not obvious because they've been changed by geologic forces, eroded away, covered by sediment, or are submerged underwater. Scientists have identified each crater by using several different kinds of clues, such as pieces of iron-rich meteorite or satellite imaging.

MATERIALS

- Cardboard box, larger than a shoebox and fairly deep (Something like a small, movable box would be perfect.)
- About 10 pounds of flour
- Ruler
- Three different-size objects that are nearly spherical, such as a rubber bouncing ball, a baseball, and a piece of roundish fruit (bigger than the baseball, such as a grapefruit). Smaller objects, such as marbles and beads, will not work well for this activity.

- Find a surface where the box is stable and you can comfortably drop the round objects from about 2 feet (0.6 m) above the surface of the flour (once it is poured into the box). This workspace should also be easy to clean; some of the flour might end up on the surrounding surfaces.

PREPARATION

- Slowly pour flour into the cardboard box. Shift the box from side to side to evenly distribute the flour. It should be a depth of at least 2 inches (5 cm) in your box. If there's not enough flour, you can either transfer the flour to a smaller box or add another bag of flour.

PROCEDURE

- Using a ruler, measure the diameter of one of your nearly spherical objects. These objects will serve as "meteorites" in this activity. *What is the diameter of this object?*

- Drop the meteorite from about 2 feet (0.6 m) above the flour into the flour by holding the object out at arm's length over the box and letting go. *What happened when the meteorite hit the surface?*

- Carefully remove the object from the flour—without disturbing the "crater" left behind. Then drop the same object into the flour box two more times, each time in a different spot in the box. Be sure to drop all of your meteorites the same way and from the same height.

- Now carefully measure the diameter of each of the three craters. (Do this by measuring the distance across the center of the depression in the flour.) *What are the diameters of the craters? How do they compare with the diameter of the meteorite itself?*

- Prepare your box for the next meteorite by shaking the box from side to side until the flour is smooth and level.
- Measure the diameter of one of your other meteorites and drop it into the box of flour just as you did with the first meteorite. Drop it two more times to make three separate craters, as you did before (from the same height), and measure the diameter of each crater. *What are the diameters of the craters? How do they compare with the diameter of the meteorite itself?*
- Again, prepare your box for the last meteorite by shaking the box from side to side to make the flour smooth and level.
- Follow the same steps with the last meteorite by dropping it into the flour box three times and measuring the diameter of the craters, just as you did with the other two meteorites. Measure the diameter of these craters. *What are the diameters of the craters? How do they compare with the diameter of the meteorite itself?*
- *What size craters did the smallest objects make? What size craters did the biggest ones make? In general, how do you think the diameter of a meteorite is related to the diameter of the crater it makes on impact?*

EXTRA

Do this activity again but this time pick only one of your "meteorites" and try dropping it from different heights. *Does the height at which the meteorite is dropped affect the crater that forms?*

EXTRA

You could graph your results from this activity, plotting the average diameter of the crater on one axis and the diameter of the meteorite on the other axis. *Do you notice any pattern between the size of the crater and the size of the meteorite, based on your graph?*

SCIENCE FAIR IDEA

Repeat this activity using objects that are of similar sizes, but different in weight. *How is a crater affected by the weight of the meteorite?* Make a graph of your results.

OBSERVATIONS AND RESULTS

Were the craters always larger than the meteorites that formed them? Did the smallest objects make the smallest craters and the largest ones make the largest craters?

When a real meteorite collides with a planet or a moon, it not only creates a crater but often explodes on impact. In this activity, there was not anywhere near the amount of force that is needed to explode the meteorite but you still should have seen that it hits the flour hard enough to make a crater that is slightly larger in diameter than the meteorite itself. The smallest craters should have been made by the smallest object, and the largest craters should have been made by the largest object.

An exception to this may have been if you used a lightweight and/or hollow object that did not sink at least halfway into the layer of flour—if this happened, a relatively smaller crater may have formed because the part of the object that hit the flour was smaller in diameter than the object's actual, widest diameter. If this happened, you could try dropping all of the objects from a greater height to give them more speed on impact.

CLEANUP

Clean any flour from your workspace. If the cardboard box is clean enough after you discard the flour from it, recycle the box. Return all other objects to the places where you found them.

Measure the Pressure

SUNNY OR STORMY? LEARN HOW A HOMEMADE BAROMETER MIGHT BE ABLE TO HELP YOU PREDICT THE WEATHER.

PROJECT TIME

45 to 60 minutes

KEY CONCEPTS

Physics
Weather
Air
Atmospheric pressure

Have you ever watched a weather forecast on TV? If so, you might have noticed the letters "H" and "L" moving around on the weather map. They are often referred to as zones of "high pressure" (H) and "low pressure" (L). The pressure they are talking about is the atmospheric pressure. Changes in air pressure can forecast short-term changes in the weather. But how do you know if the air pressure changes—or if it is high or low? Scientists have developed an instrument called a barometer that can measure atmospheric pressure. In this activity, you will find out how a barometer works by building one yourself!

BACKGROUND

Earth is surrounded by a layer of air called the atmosphere. The atmosphere is about 300 miles (483 km) thick, and most of it is within 10 miles (16 km) of Earth's surface. The air is made up of gas molecules such as nitrogen, oxygen, and carbon dioxide. All of these molecules press down on Earth, and this pressure is called air pressure or atmospheric pressure.

You might be surprised to learn that air has a weight, although you don't feel it. We have evolved so that the pressure inside our body matches the external atmospheric pressure. You are, however, able to feel changes in atmospheric pressure. For example, if you fly in an airplane, your ears might start to hurt. This is because at high altitude your internal pressure is greater than the external pressure. Only when you balance both pressures by making your ears "pop" will they stop hurting.

The atmospheric pressure is highest at sea level and decreases when you move up into the atmosphere. At sea level, normal air pressure ranges from 800 to 1,050 millibars. On top of Mount Everest, the air pressure is about 30 percent less! Air pressure changes with altitude and with temperature. Warm air, which is less dense, rises and results in lower air pressure. When it rises it cools down and turns into water vapor, which then condenses into liquid. This leads to cloud formation and rain. Low pressure, therefore, is generally associated with cloudy and rainy weather. On the other hand, high-density cold air results in increased air pressure. As cold air sinks, it dries, causing warm and dry weather conditions in high pressure zones.

Now that we know how air pressure can help predict the weather, we still need to find out how to measure air pressure. To do this, scientists have developed the barometer. Many different types of barometers exist. One of them is the water-based barometer: water is sealed in a glass container with a narrow spout connecting the outside to the inside. When the outside air pressure increases it pushes onto the water in the spout, resulting in a water level decrease. When less air (lower air pressure) pushes onto the water inside the spout the water level rises. In this activity, you will build a variation of this barometer. Do you think it can predict tomorrow's weather?

MATERIALS

- Heat-resistant glass jar
- Rubber band that fits around the mouth of the glass jar
- Balloon
- Wooden skewer
- Scissors

- Tape
- Paper
- Pen
- Three bowls
- Tap water (hot and cold)
- Ice cubes
- Paper towels

PREPARATION

- Cut off and discard the opening of the balloon, then stretch the remaining balloon piece around the opening of the glass jar, wrapping it tightly.
- Secure the balloon to the jar with the rubber band.
- Tape the end of the wooden skewer flat onto the top of the balloon with the pointy tip pointing away from the jar. You have now completed making your barometer.
- Fill the first bowl with room temperature water, the second bowl with hot tap water (be careful with hot water and have an adult help if necessary), and the third bowl with ice water.
- Place your barometer close to a wall with the skewer parallel to the wall.
- Tape a piece of paper behind the barometer on the wall.
- Draw a line on the paper that lines up with the tip of the skewer.
- Mark the position where the barometer is placed. You will have to place it on the same spot later during the activity.

PROCEDURE

- Place the glass jar of your barometer in the bowl with the hot water. The water should not reach beyond the rubber band. *What do you observe about the barometer? Do you notice any changes? If yes, what kind of changes?*

- After one minute, take the barometer out of the water, quickly dry it off, and place it at the same spot close to the wall where you have attached the paper. Make another line on the paper so that it lines up with the pointy tip of the skewer. *Where does the tip point now? How did the barometer change?*
- Place the barometer into the bowl with room temperature water. Again, observe what happens. Keep it in the bowl for about 10 minutes. *After 10 minutes, how does the barometer look different from before?*
- After 10 minutes, put the barometer back at its spot by the wall and make another line on the paper where the skewer is pointing. *What do you notice this time?*
- Next, place the barometer into the bowl with ice water for about one minute. *What happens to the barometer in the ice water? What differences do you notice?*
- Again, place the barometer in front of your paper chart, and make another line where the skewer tip is pointing. *Can you explain your observations?*
- Place the barometer into the bowl with room temperature water and repeat the steps once more. *Where does the skewer tip point to this time? Why?*

EXTRA

Repeat your tests with different water temperatures. *How do your results change with other water temperatures?*

EXTRA

Take your barometer outside to measure atmospheric pressure. Draw a scale that looks like a ruler on another piece of paper. Then place the chart next to the barometer so that the middle line lines up with the skewer tip. Leave the barometer outside for several days, and every day mark where the skewer is pointing. *What does it mean when the tip is moving up or down on your scale?*

OBSERVATIONS AND RESULTS

Were you able to measure the change in air pressure with your barometer? In this activity, you actually measured the air pressure changes inside the glass jar—not the external atmospheric pressure change. As we cannot change the external atmospheric pressure ourselves we changed the air pressure inside the glass jar to demonstrate how the balloon barometer works.

When you seal the glass jar with the balloon, there is a certain air pressure inside the glass jar. The tip of the wooden skewer should be pretty level with the top of the glass jar as the balloon surface is very flat. When you put the glass jar in the hot water, however, the air inside the jar heats up and expands. As the jar is sealed with the balloon it has nowhere to escape—and the balloon starts to bulge. This is because the air pressure inside the glass increases and is pressing against the balloon. As a result, the wooden skewer points downwards. The line that you made should be below the original line.

When you put the balloon into room temperature water, the air is cooling back to room temperature where it was before, so your balloon should become flat again. At the same time, the wooden skewer should line up with your original line. The opposite happens when you put the glass jar in ice water. As the air inside the jar cools down even more, it contracts and pulls the balloon into the jar. This makes the wooden skewer point upwards. The line that you made for ice water should be above your original line. Back in room temperature water, the balloon and the wooden skewer should get back to their original positions.

If you measure external air pressure with your barometer, the reverse will happen: the skewer will point upwards with rising atmospheric pressure as the air outside the jar pushes onto the balloon. With falling atmospheric pressure, the wooden skewer will point downwards. Marking the location of the wooden skewer every day can tell you what kind of weather to expect!

CLEANUP

Use the water in your barometer to water some plants. Throw away any trash. Return all other materials to the places where you found them.

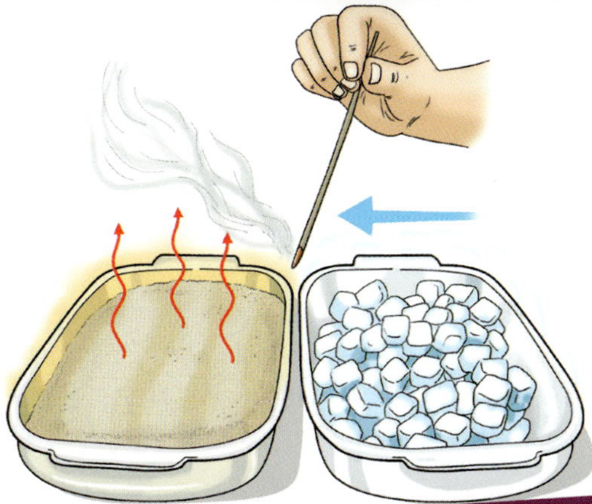

Create a Sea Breeze

LEARNING ABOUT TEMPERATURE AND PRESSURE IS A BREEZE WHEN YOU APPLY A LITTLE PHYSICS TO YOUR OWN HOMEMADE BEACH!

A hot summer day is the perfect time to go to the beach and cool down in the brisk ocean water. But it's not only the water that has a cooling effect at the beach. Have you ever noticed there always seems to be a cool breeze blowing from the ocean to the shore? Where does the wind come from? In this activity, you will build a model of the ocean and beach to find out—so you will know why the sea breeze is blowing!

PROJECT TIME

60 minutes

KEY CONCEPTS

Physics
Air pressure
Density
Heat capacity
Temperature

BACKGROUND

If you were asked to describe wind, you might say it is the movement of air from one place to another. But have you ever thought about what makes the air move? It all starts with the sun heating Earth's surface. Not every surface heats up the same way: Some surfaces take longer to warm whereas others get hot right away. This is because different materials have different heat capacities. Surfaces with a low heat capacity will heat and cool quickly but those with a high heat capacity take much longer to do so. Water has a higher heat capacity, which means it takes a lot of heat to increase its temperature. Beach sand, on the other hand, has a lower heat capacity and warms pretty quickly in the sun, which you might have noticed when walking on the sand barefoot. But what do water and sand temperatures have to do with wind?

The air above the beach warms because of the hot sand. As the air heats up, it becomes less dense and rises. The expanding air results in a decrease in air pressure (how much pressure the air exerts on Earth's surface at a certain location). Warm air rises, causing low air pressure whereas cold, dense air results in high air pressure. Above the ocean the air cools down due to the colder temperatures of the ocean water. This leads to the air becoming denser and a local high air pressure zone.

The difference in air pressures above the beach and ocean causes the air movement we perceive as wind. In an attempt to balance the pressures in both areas, the air from the high-pressure zone (the ocean) will rush to the low-pressure zone (the beach) and replace the rising air. The outcome of this airflow is the cool breeze that blows from the ocean toward the beach. The strength of the sea breeze depends on the difference in temperature between the land and ocean. Higher temperature gradients lead to stronger winds. During the night this sea breeze turns into a land breeze, because after sunset the sand will cool down much faster than the water due to its lower heat capacity. Once the sand becomes colder than the ocean the airflow pattern reverses. Want to see for yourself? Create your own sea breeze in this activity—no real ocean or beach required!

MATERIALS

- Two baking dishes of the same size, preferably glass
- Sand (enough to fill one baking dish)
- Workspace that has no wind
- Pot
- Stove
- Incense stick

- Lighter
- Adult helper
- Ice water
- Hot water
- Freezer
- Large cardboard box (optional, if needed to shield any breezes)
- Thermometer (optional)

PREPARATION

- Put the sand in a pot, and with the help of an adult heat it on the stove until it is very hot to the touch (about 158 to 176°F [70 to 80°C])

PROCEDURE

- Place the hot sand in one of the baking dishes. Carefully feel the sand with your fingers. *How hot is the sand?*
- Fill the other baking dish with ice water. Feel the temperature of the water with your fingers. *How cold is the water?*
- Put the dishes next to each other at a place where there are no external winds. You can use the cardboard box to shield the dishes from any airflow.
- Make sure there is no air movement above the dishes. Then, with the help of an adult, light the incense stick. Once it makes lots of smoke, hold it in between the sand and ice water dishes. Observe the smoke's movement above the dishes for about one minute. *Where does the smoke go? Is it blowing toward the sand dish or the ice water dish? Can you explain why?*
- Empty both dishes and let the sand cool to room temperature. Then, fill one dish with the room-temperature sand and the other dish with room-temperature water. Feel the temperature of the sand and the water with your finger. *Are the temperatures of both very different?*

- Place both dishes next to each other in your wind-protected area. Use the same incense stick and hold it between both dishes. *Where does the smoke go this time?*
- Next, put the dish with the sand into the freezer for about 10 to 15 minutes. While the sand is cooling down, heat water on the stove to about 158 to 176°F (70 to 80°C). Replace the room-temperature water with the hot water in the second dish.
- Once the sand is cooled and the water is heated, place the dishes next to each other again in your wind-protected area.
- Hold the incense stick between the dishes again and observe which direction the smoke is moving. *Does it move toward the sand or the water? Can you explain why?*

EXTRA

Repeat this activity with other temperature differences between the water and the sand. *How high does the temperature difference need to be to observe a sea or land breeze?*

SCIENCE FAIR IDEA

Would your observations be different at a pebble beach? Replace the sand with pebbles and repeat the activity to find out!

SCIENCE FAIR IDEA

How would your results be affected by external winds? Remove your wind shield or repeat the activity in a windy area to see how your air flow is affected.

OBSERVATIONS AND RESULTS

Can you see how you created a little beach model? The dish with sand represented the beach and the one with water represented the ocean. You heated the sand on the stove like the sun heats the sand at the beach. When the sand is much warmer than the water you should have seen the smoke of the incense stick steadily flow from the water to the sand dish. This airflow is caused by the temperature differences between the sand and water. The air above the hot sand heats up and starts to rise as it becomes less dense, creating a local low air pressure zone. Above the cold-water dish, the air cools and becomes denser, leading to a local high air pressure zone. To balance out the air pressure differences the air moves from the high air pressure zone (the water dish) to the low air pressure zone (the sand dish), and the smoke is carried along with it.

If you remove the temperature difference, as you did in your second test, no airflow will be created. The air above the sand and water have the same air pressure, which is why you shouldn't have seen the smoke move into a particular direction unless you had an external airflow. In the last test, you created a land breeze from the sand to the water, which happens at night when the air flows from beach to ocean. In this case, the sand is cooler than the water, which reverses the airflow due to the changes in air pressure above both the beach and ocean.

CLEANUP

Make sure to fully extinguish the incense stick. Dispose of the sand outside if you have permission or in the trash. Clean your baking dishes with warm water and soap.

Warm in the Sun

COOL IN THE SHADE? LEARN ABOUT HOW DIFFERENT MATERIALS ABSORB AND REFLECT SOLAR RADIATION TO MAKE OUTDOOR ENVIRONMENTS SO VARIABLE.

Have you ever walked across a large parking lot on a sunny summer day and felt like you were roasting? That's because the asphalt gets really hot in the sun! Streets, buildings, and parking lots can get so hot, they raise the average temperature of urban areas by a few degrees relative to surrounding rural areas. Do you think natural materials also heat up in the sun—or only human-created materials? Try this activity to find out!

PROJECT TIME

20 minutes

KEY CONCEPTS

Physics
Heat
Sun
Materials

BACKGROUND

The sun is some 93 million miles (150 million km) from Earth. But because its surface is so hot (about 10,000°F [5,538°C]), we on Earth benefit from the light it radiates—without it, humans could not survive. The light radiated by our sun carries energy, part of which gets absorbed and transformed into heat when it reaches a surface. That is why places in the sun feel warmer than those in the shade. Another part of the light is reflected. That part makes surfaces look lighter and brighter in sunlight.

Earth is covered with water, soil, rocks, snow, and ice as well as a wealth of human-made materials, such as concrete. When these surfaces are exposed to the sun, they warm up. This process takes time—and some materials, such as water, need more light from the sun to warm up a few degrees than do others, such as soil or rocks. Once warmed, these surfaces release heat, which warms the air above them. This process is also gradual, and some materials including water hold onto heat longer than do others, such as soil or rocks. This difference generally makes the temperature of coastal climates more moderate than those inland. But other natural factors, such as trees and plants, moderate the air temperature too. Plants absorb water from the ground and transpire (release water into the air), which works like a natural air conditioner.

Many factors determine how much materials heat up in the sun and how they release this energy. In this activity, you will test some materials yourself.

MATERIALS

- Sunny day
- Metal wire (about 3.2 feet [1 m])
- Partially shaded area outside with at least two different ground covers (dirt, grass, rocks, concrete, etcetera)

PREPARATION

- Find the line on the ground where an area in the sun meets an area in the shade. Place your wire on this line.

PROCEDURE

- Take a big step into the shade. *What covers Earth's surface in this particular spot?*

- Find a spot in the sun where the surface is covered with the same material. Choose one that is at least one big step away from a shaded area. *Does the spot in the sun look different compared with the spot in the shade? Why do you think this is the case?*

- *Do you think these spots will feel different too?*

- Carefully touch both places with your hand and compare how they feel. *Does one feel warmer than the other? If so, which one feels warmer? Why would this be the case?*

- If it's possible and you have permission, dig up about inch of surface material in each place and feel the layer beneath it. This may not work in some locations or with some materials. (For example, you shouldn't dig holes in a public park, and you can't dig holes in concrete.)

- *For the material in the sun, do the surface and the layer underneath feel similar or is one warmer? Do you note the same findings for the material in the shade?*

- Find a different material. *Can you find one spot where the surface is covered with this material in the shade and one in the sun? If so, do these spots look different? Do they also feel different?*

- If you can, repeat this investigation with a third material: include water, soil, or another natural material!

- *Do you see a pattern? Why would this happen?*

- Go back to the wire you placed at the edge of the shade. *Can you see how the edge of the shade moved?* If it hasn't moved, wait 10 minutes and then come back. *Why would the edge of the shade move?*

- Depending on which way the shadow moved, you now have a new area in the shade that was recently in the sun or vice versa. Investigate how this area looks and feels, compared with areas that have been in the sun or the shade for a long time. *Do you notice any differences?*

EXTRA

If you found a pattern, test it on more materials the next time you are in the park, at a playground, or at an outdoor sports field—or repeat the procedure inside. *Do the same differences occur? Why or why not?*

SCIENCE FAIR IDEA

Place cups of water, dirt, and/or pebbles in the shade and identical cups in the sun. Investigate how the materials in the sun look and feel compared with those in the shade. Note you cannot compare the temperatures of two different materials (for example, rocks and dirt) by feeling them with your hands, because different materials at the same temperature can feel very different. You can compare the same materials placed in the sun and shade or use an infrared thermometer to compare the surface temperature of different materials.

EXTRA

Try to find out if there is a correlation between color and how much a material warms up when placed in the sun. *Do dark soil or pebbles warm up more than, less than, or around the same amount as light-colored soil or pebbles? What about darker- and lighter-colored cement?*

EXTRA

Find information on the urban heat island effect and how cities try to combat it. *Can you design a small-scale test to investigate if these ideas work?*

OBSERVATIONS AND RESULTS

Did the materials look brighter and lighter in the sun? Did they also feel warmer?

Our sun radiates light; some of it reaches Earth. When this light hits soil, pebbles, grass, cement, or any other material that covers our planet, part of it bounces back. This reflected light makes the material look brighter and lighter. The other part of the light gets absorbed or sucked into the material and warms its surface. That is why materials in the sun feel warmer when you touch them. Warm materials heat the air surrounding them, which starts to feel warmer too.

Maybe you noticed a material became just a little bit warmer after the sun initially shone on it—and after a while became much warmer. Warming is a process that takes time—just as when a material in the sun moves to the shade it cools down gradually.

Some materials, such as cement, dark rocks, or even sand, can get very hot when left in the sun for a while. Water warms too, but it needs to stay in the sun longer to heat up by the same amount; it also cools down more slowly than soil or cement.

CLEANUP

Return the metal wire to where you found it or dispose of it if you cannot reuse it.

It's About Time... to Make a Sundial!

CLOCK TIME: MEASURE THE MINUTES AS THE SUN SHIFTS OVERHEAD. ALL YOU NEED IS SOME SUNSHINE, SIMPLE HOUSEHOLD ITEMS—AND SOME TIME!

PROJECT TIME

3 hours

KEY CONCEPTS

Physics
Earth's rotation
Time
Sundial

Have you ever watched a movie set in an earlier era, and when a character asks what time it is, the other characters, who don't have watches or cell phones, all look at the sky? They are not looking at a giant digital clock above, they are using the position of the sun in the sky to tell time, as people have done for generations. The oldest known instrument for telling time, the sundial, allows us to track the position of the sun more accurately. Up until the early 19th century, sundials were the main instrument people used to tell time. If they are correctly placed, sundials can be used to accurately tell time down to the minute! In this activity, you will be making your own sundial and using your body to track the movement of the sun across the sky!

BACKGROUND

For millennia, people have used sundials to tell the time of day based on the apparent position of the sun in the sky. There are many types of sundials, but in general, each consists of a gnomon, a thin rod that casts a shadow onto a dial, and a flat plate or platform. The apparent movement of the sun across the sky is the result of Earth's rotation on its axis. As our planet spins, the sun appears to move across the sky—but really, we're the ones who are moving!

As the sun's position changes in our sky, the shadow it casts will align with lines marking each hour indicating the time of day. The accuracy of a sundial is affected by a number of factors, including the fact that the angle of Earth's rotation isn't perfectly perpendicular, and Earth isn't perfectly round. As a result, corrections have to be made to sundials to account for these changes.

For this activity, you'll be making a simple sundial (using a clock to help you!) as well as tracking the position of the sun by observing your shadow.

MATERIALS

- Sidewalk chalk
- Tape measure or yardstick
- Pen or pencil
- Large concrete space with no shadows
- Clock
- Paper plate
- Plastic straw
- Ruler
- Markers or crayons
- Paperweight or a few small stones
- Adult helper
- Sunny weather

PREPARATION

- This activity works best if you start early in the day, so you have a few hours of daylight to do your testing. We recommend starting at 9 a.m. and testing off and on until noon—or starting at noon and testing off and on until 3 p.m.

PROCEDURE

- Start by choosing a place where you will always stand during this activity. Make sure it is in the middle of the open concrete space, with no shadows nearby. Mark this space by using your chalk to outline your shoes.

- Stand in your chosen spot and have your helper use the sidewalk chalk to trace the outline of your shadow on the concrete.

- Use your chalk to write the time at the top of your shadow.

- Repeat these steps every 30 minutes, each time marking the time of day at the top of your shadow.

- While you are waiting to trace your shadow use a pencil or pen to carefully poke a hole through the center of your paper plate.

- Check the time, round up to the nearest hour, and write this number at the very edge of your plate with a crayon or marker. For example, if the clock says 9:45 a.m., write "10" on the plate. Use your ruler to draw a straight line from the number you wrote to the hole in the center of the plate.

- Wait until the clock reads the hour that you wrote before proceeding to the next step (for example, if you started at 9:45 a.m., you would wait 15 minutes until the clock reads 10 a.m.).

- Take your plate and plastic straw outside. Put the plate on the ground and poke the straw through the hole you made. Slant the straw slightly toward the line you drew.

- Carefully rotate the plate so the shadow of the straw lines up with the line you drew. *Do you think the shadow will stay in the same place all day? Why or why not?*

- Place the paperweight or stones on the very edges of the plate to hold it in place.
- Every hour, check your sundial and the position of the shadow on your plate. If you started at 10 a.m., note the position of the shadow at 11 a.m. and write "11" on the edge of the plate where this shadow falls. Each time you check the sundial, write the hour on the edge of the plate where the shadow falls. *Why do think the shadow is moving? What does your sundial remind you of?*
- After several hours of tracing your own shadow, observe the positions of each tracing. *Did your shadow move during the day? What else changed about your shadow with each tracing? What do you think caused these changes?*

OBSERVATIONS AND RESULTS

In this activity, you observed the movement and changes in shadows over the course of the day. In the case of your own shadow, the pattern you noticed depends on the time of day you started. If you started this activity in the morning, you should have observed your shadow started out long and by the middle of the day it looked much shorter. If you started in the middle of the day, you noticed the opposite—your shadow started off shorter and grew longer over the course of the afternoon. Regardless of when you started, however, you should have noticed the position of your shadow changed over time. As it got later your shadow moved in a clockwise direction from the first outline you drew—as long as you were completing this activity in the Northern Hemisphere!

For your sundial you should have noticed something similar. At each hour, the shadow of the straw was in a different position, each time moving clockwise from the start position. After a few hours you should have noticed the sundial looks like the face of a clock with the numbers evenly spaced out around the plate.

The reason for your shadow's change in shape and position has to do with Earth's rotation on its axis. As Earth spins, the sun appears to move across the sky. The sun is highest in the sky at noon or midday, and at this point it casts its shortest shadow because it is most directly above us in the sky. In the morning and later in the afternoon, the sun is more off-center and therefore casts a longer shadow.

The position of the shadow also changes as the sun appears to move across our sky. You can see something similar if you shine a flashlight on your hand, and then move the flashlight. As you move the light, the position of your hand's shadow will change with the movement of the flashlight. The position of the sun in our sky is dictated by the speed of Earth's rotation—it turns on its axis at a speed of 460 meters per second, or approximately 1,000 miles per hour! When Earth rotates 15 degrees on its axis, it's just as though the sun has moved 15 degrees across our sky. Due to this movement (depending on where you live), the shadow cast by the sun moves about 15 degrees each hour so that over 24 hours it travels a full 360 degrees around your sundial.

CLEANUP

Return all the materials you used to the places where you found them.

THE SCIENTIFIC METHOD

The scientific method helps scientists—and students—gather facts to prove whether an idea is true. Using this method, scientists come up with ideas and then test those ideas by observing facts and drawing conclusions. You can use the scientific method to develop and test your own ideas!

Question: What do you want to learn? What problem needs to be solved? Be as specific as possible.
Research: Learn more about your topic and refine your question.
Hypothesis: Form an educated guess about what you think will answer your question. This allows you to make a prediction you can test.
Experiment: Create a test to learn if your hypothesis is correct. Limit the number of variables, or elements of the experiment that could change.
Analysis: Record your observations about the progress and results of your experiment. Then analyze your data to understand what it means.
Conclusion: Review all your data. Did the results of the experiment match the prediction? If so, your hypothesis was correct. If not, your hypothesis may need to be changed.

GLOSSARY

air pressure: The weight of air molecules pushing down on Earth.
atmosphere: The mixture of gases that surrounds the Earth.
axis: An invisible line an object rotates around.
barometer: A device that measures atmospheric pressure.
black hole: An object in space that almost nothing can escape.
crater: A bowl-shaped or hollowed-out area.
friction: Resistance when one object rubs against another.
gravity: A force that brings two objects together.
heat capacity: How much energy is required to raise the temperature of a material.
hemisphere: One-half of Earth.
meteor: A piece of rock or metal that burns up in Earth's atmosphere, causing a streak of light.
radiate: To emit rays of light or heat.
rotate: The circular motion of an object around its center.
slope: The steepness and direction of a line.

ADDITIONAL RESOURCES

Books

Clarke, Izzie. *Space*. New York, NY: Kingfisher, 2023.

Denecke, Edward J. *Painless Earth Science*. New York, NY: Kaplan, Inc., 2021.

Lynch, Dan R. *The Earth Book for Kids: An Introduction to Earth Science*. Cambridge, MN: Adventure Publications, 2022.

Websites

Discovery Education
sciencefaircentral.com

Exploratorium
www.exploratorium.edu/explore/activities

Science Buddies
www.sciencebuddies.org/science-fair-projects/project-ideas/list

Science Fun for Everyone!
www.sciencefun.org/?s=science+fair

Videos

The Earth
https://wqed.pbslearningmedia.org/resource/the-earth-science-trek/the-earth-science-trek/

"Space"
https://wqed.pbslearningmedia.org/resource/space-super-science-joe-video/super-science-show-with-joe/

Publisher's note to educators and parents: Our editors have carefully reviewed these websites to ensure that they are suitable for students. Many websites change frequently, however, and we cannot guarantee that a site's future contents will continue to meet our high standards of quality and educational value. Be advised that students should be closely supervised whenever they access the internet.

INDEX

A

air, 39, 40, 41, 42, 43, 45, 46, 47, 48, 49
astronomy, 15, 16, 17, 18, 19, 20, 21, 22, 23, 24, 25, 33, 34, 35, 36, 37
atmosphere, 10, 34, 39, 40, 41, 42, 43

B

barometer, 39, 40, 41, 42, 43
black hole, 5, 6, 7, 8

C

crater, 33, 34, 35, 36, 37

D

density, 9, 10, 11, 12, 13, 45, 46, 47, 48, 49

E

Einstein, Albert, 5, 6

F

force, 5, 6, 7, 8, 27, 28, 29, 30, 31
friction, 27, 28, 29, 30, 31, 34

G

general relativity, 5, 6
geometry, 33, 34, 35, 36, 37
gravity, 5, 6, 7, 8, 27, 28, 29, 30, 31

L

landslide, 27, 28, 29, 30, 31
light, 5, 6, 7, 8, 21, 22, 23, 24, 25
liquid, 9, 10, 11, 12, 13

M

mass, 5, 6, 7, 8
measurement, 15, 16, 17, 18, 19, 20
meteor/meteorite, 33, 34, 35, 36, 37

N

Newton, Isaac, 6

O

orbit, 21, 22, 23, 24, 25

P

parallax, 15, 16, 17, 18, 19, 20
physics, 5, 6, 7, 8, 15, 16, 17, 18, 19, 20, 33, 34, 35, 36, 37, 39, 40, 41, 42, 43, 45, 46, 47, 48, 49, 51, 52, 53, 54, 55, 56, 57, 58, 59, 60
pressure, 9, 10, 11, 12, 13, 39, 40, 41, 42, 43, 45, 46, 47, 48, 49

R

rotation, 56, 57, 58, 59, 60

S

science fair ideas, 9, 27, 33, 45, 51
scientific method, 61
seasons, 21, 22, 23, 24, 25
slope, 27, 28, 29, 30, 31
solar system, 6
space-time, 5, 6, 7, 8
star, 6, 8, 15, 16, 17, 18, 19, 20
sun, 21, 22, 23, 24, 25, 46, 51, 52, 53, 54, 55, 56, 57, 58, 59, 60

T

temperature, 45, 46, 47, 48, 49, 51, 52, 53, 54, 55
time, 56, 57, 58, 59, 60

W

waves, 9, 10, 11, 12, 13
weather, 39, 40, 41, 42, 43
wind, 45, 46, 47, 48, 49